BIOLOGY PRACTICAL GUIDE 3
CELLS, TISSUES, AND ORGANISMS IN RELATION TO WATER

Revised Nuffield Advanced Science
Published for the Nuffield–Chelsea Curriculum Trust
by Longman Group Limited

Longman Group UK Limited
Longman House, Burnt Mill, Harlow, Essex, CM20 2JE, England
and Associated Companies throughout the world.

First published 1970
Revised edition first published 1985
Fourth impression 1990
Copyright © 1970, 1985, The Nuffield-Chelsea Curriculum Trust

Design and art direction by Ivan Dodd
Illustrations by Oxford Illustrators

Set in Times Roman and Univers
Produced by Longman Group (FE) Ltd
Printed in Hong Kong

ISBN 0 582 35429 3

All rights reserved. No part of this publication may be reproduced, stored in a retrieval system, or transmitted in any form or by any means, electronic, mechanical, photocopying, recording, or otherwise without either the prior written permission of the Publishers or a licence permitting restricted copying issued by the Copying Licensing Agency Ltd, 33-34 Alfred Place, London, WC1E 7DP.

K.C.F.E. Library	
Class No.	574.072
Acc. No.	00036846
Date Rec	19/06/91
Order No.	B 322542

Cover photograph
A light-micrograph of *Paramecium* ($\times 1500$). See investigation 9E, 'The action of a contractile vacuole'.
Photograph, Dr David J. Patterson, Department of Zoology, University of Bristol.

CONTENTS

Introduction page 1

Chapter 8 **The plant and water 4**
- 8A Is the water content of a plant constant or does it fluctuate? 4
- 8B Do roots affect the rate of water input into land plants? 5
- 8C Is the rate of water input (and output) affected by environmental conditions around the aerial portions of the plant? 6
- 8D What is the role of the leaf in the water vapour loss from a cut shoot? 8
- 8E Translocation paths for water in plant stems 11
- 8F The anatomy of a stem and how water moves up 12
- 8G The xylem pathway 15
- 8Ha Water movement by imbibition into a non-living physical system and into live seeds 17
- 8Hb Increase in the mass of seeds in different soils and known solutions 18
- 8I A demonstration of the pressure produced during imbibition 19
- 8J Osmotic water movement into a non-living system and into living plant tissue 20
- 8K A demonstration of the osmotic input into seedlings, producing root pressure so that the guttation droplets are extruded at special points on the leaves 22
- 8L Water movement against gravity in a physical system and in plants 22
- 8M A demonstration of the mechanical support function of turgor pressure produced by water input into leaf cells 25

Chapter 9 **The cell and water 26**
- 9A Measurement of the water potential of cell sap by plasmolysis 26
- 9B Diffusion of substances against gravity in a gel 28
- 9C Movement through a differentially permeable membrane 28
- 9D The effect, on isolated animal cells, of altering the composition of the external medium 29
- 9E The action of a contractile vacuole 30

Chapter 10 **Control by the organism 34**
- 10A The relation of the urinary system of a mammal to other systems of the body 34
- 10B Injection of the arterial blood system in the kidney 41
- 10C The histological structure of the nephron 44
- 10D Determination of chloride content of urine collected in different circumstances of salt intake 53

SAFETY

In these *Practical guides*, we have used the internationally accepted signs given below to show when you should pay special attention to safety.

 highly flammable

 explosive

 toxic

 corrosive

 radioactive

 take care! (general warning)

 risk of electric shock

 naked flames prohibited

 wear eye protection

 wear hand protection

INTRODUCTION

The practical investigations in this *Guide* relate to the topics covered in *Study guide I*, Part Two, 'Control and co-ordination in organisms', Chapters 8–10. Cross references to the *Study guide* are given.

Chapter 8 THE PLANT AND WATER

Investigation 8A Is the water content of a plant constant or does it fluctuate? (*Study guide*. Study item 8.11 '*Vegetable staticks*'.)
Fluctuations in the water content of seedlings are investigated by weighing.

Investigation 8B Do roots affect the rate of water input into land plants? (*Study guide*. Study item 8.11 '*Vegetable staticks*'.)
The rate of water input into seedlings with and without roots is compared by measuring loss in mass.

Investigation 8C Is the rate of water input (and output) affected by environmental conditions around the aerial portions of the plant? (*Study guide*. Study item 8.11 '*Vegetable staticks*'.)
A potometer is used to measure the rate of water input into a plant.

Investigation 8D What is the role of the leaf in the water vapour loss from a cut shoot? (*Study guide*. Study item 8.11 '*Vegetable staticks*'.)
A potometer and a microbalance are used to investigate the role of leaves in the loss of water vapour.

Investigation 8E Translocation paths for water in plant stems. (*Study guide* 8.3 'The cohesion theory of water movement in the plant'; Study item 8.31 'The rise of water in a plant'.)

Investigation 8F The anatomy of a stem and how water moves up. (*Study guide* 8.3 'The cohesion theory of water movement in the plant'; Study item 8.31 'The rise of water in a plant'.)

Investigation 8G The xylem pathway
This investigation provides firsthand evidence by dissection of the xylem pathway in a plant stem.

Investigation 8Ha Water movement by imbibition into a non-living physical system and into live seeds. (*Study guide* 8.2 'Imbibition and osmosis in seeds and seedlings'.)

Investigation 8Hb Increase in the mass of seeds in different soils and known solutions. (*Study guide* 8.2 'Imbibition and osmosis in seeds and seedlings'.)

Investigation 8I A demonstration of the pressure produced during imbibition. (*Study guide* 8.2 'Imbibition and osmosis in seeds and seedlings'.)
A lever system is used to demonstrate the pressure produced during imbibition.

Investigation 8J Osmotic water movement into a non-living system and into living plant tissue. (*Study guide* 8.2 'Imbibition and osmosis in seeds and seedlings'.)
An osmometer is used to investigate osmotic water movement.

Investigation 8K A demonstration of the osmotic input into seedlings, producing root pressure so that guttation droplets are extruded at special points on the leaves. (*Study guide* 8.2 'Imbibition and osmosis in seeds and seedlings'.)

Investigation 8L Water movement against gravity in a physical system and in plants. (*Study guide* 8.3 'The cohesion theory of water movement in the plant'.)
An atmometer is used to study water movement.

Investigation 8M A demonstration of the mechanical support function of turgor pressure produced by water input into leaf cells. (*Study guide* 8.4 'What role does water play in the life of a plant?')

Chapter 9 THE CELL AND WATER

Investigation 9A Measurement of the water potential of cell sap by plasmolysis. (*Study guide* 9.1 'Water relations of a plant cell'; Study item 9.11 'Quantitative examples of water potential in plant cells'.)

Investigation 9B Diffusion of substances against gravity in a gel. (*Study guide* 9.1 'Water relations of a plant cell' – 'Diffusion', page 270.)

Investigation 9C Movement through a differentially permeable membrane. (*Study guide* 8.2 'Imbibition and osmosis in seeds and seedlings'; 9.1 'Water relations of a plant cell' – 'Dialysis', page 271; 10.6 'Kidney failure'.)

Investigation 9D The effect, on isolated animal cells, of altering the composition of the external medium. (*Study guide* 9.2 'Water relations of animal cells'; Study item 9.21 'The effect of various solutions on human red blood cells'.)

Investigation 9E The action of a contractile vacuole. (*Study guide* 9.2 'Water relations of animal cells'; Study item 9.22 'The action of the contractile vacuole complex'.)

Chapter 10 CONTROL BY THE ORGANISM

Investigation 10A The relation of the urinary system of a mammal to other systems of the body. (*Study guide* 10.3 'The internal environment'.)
Instructions are given for the dissection of the urinary and reproductive systems of a male and of a female mammal.

Investigation 10B Injection of the arterial blood system in the kidney. (*Study guide* 10.4 'The functioning of the kidney'.)
This technique illustrates how the blood supply to the kidney is suited to the organ's functioning.

Investigation 10C The histological structure of the nephron. (*Study guide* 10.4 'The functioning of the kidney'.)
The nephron is the unit of function in the kidney and a knowledge of its structure is essential for an understanding of its functioning.

Investigation 10D Determination of chloride content of urine collected in different circumstances of salt intake.
The chloride content of urine can be estimated by titration, and any differences may be linked with salt intake.

A note for users of this *Practical guide*

The instructions given for the investigations are intended for use as guidelines only. We hope that you will modify and extend the techniques that have been described to meet your own requirements. Other organisms should certainly be tried, depending on what is most readily available. Some of these investigations may lend themselves to further work in a Project.

It may not always be possible, for various reasons, for you to do a practical investigation. A study of data from another source is perfectly acceptable in such a case.

CHAPTER 8 **THE PLANT AND WATER**

Ever since scientific investigations into plant water relations began in the seventeenth century two questions have been asked:
1 By what mechanism does a plant obtain, translocate, and consume water?
2 What role does water play in the life of a plant?
In attempting to find an answer to the first question we must find out whether the water content of a plant is static or dynamic.

INVESTIGATION
8A Is the water content of a plant constant or does it fluctuate?

(*Study guide.* Study item 8.11 '*Vegetable staticks*'.)

Procedure
1 You need six comparable seedlings, about 15 cm high, of one of the following: French bean, broad bean, pea, or tomato. Three of these seedlings have been left unwatered for three days but should not be severely wilting; the other three have been watered daily. All six seedlings have had good illumination.
2 Cut off the seedlings at soil level, weigh each shoot separately, and record its fresh mass (FM).
3 Dry the shoots in an oven at 80 °C overnight and reweigh each shoot to obtain its dry mass (DM).
4 Calculate the mean fresh mass and mean dry mass of the seedlings and use the means for the following percentage calculation.
5 Express the water content of the shoots as a percentage of the mean dry mass.

$$\frac{(FM - DM)}{DM} \times 100 = \% \text{ mean plant water content.}$$

6 A variation of this investigation is to withhold water for three days from six seedlings, and then determine the fresh mass and dry mass of three of them as above. Water the other three seedlings and place a plastic hood over the shoots for three hours. Determine the fresh mass and dry mass for these. Calculate the percentage of plant water content for each group of three seedlings.

Question
a *What can you conclude from this investigation about the water content of plants and the factors which affect it?*

Turnover and fluctuation of the water content of plants can only come about if the rate of water input differs from the rates of water consumption by the plant or loss (output) from it. We must therefore try to find answers to the following questions:

How and where does water enter the plant?
How and where is water consumed by or lost from the plant?
How are the sites of entry and loss of water connected?

The next four investigations will provide some of the information needed to answer these questions.

INVESTIGATION
8B Do roots affect the rate of water input into land plants?

(*Study guide*. Study item 8.11 '*Vegetable staticks*'.)

Procedure

1   You need six seedlings, as in the previous investigation, all of which have been watered daily. Set the seedlings up as shown in *figure 1a* overleaf, but without the liquid paraffin, and let them acclimatize for a week (this may have been done for you).
2   For the experiment, remove the roots from three of the seedlings by cutting at the base of the stem under water (*figure 1b* overleaf). Leave the other three seedlings intact.
3   Add enough liquid paraffin to each beaker to cover the water surface. Make sure it does not come into contact with the cut stem.
4   Place each seedling assembly on a top-pan weighing machine and keep them well illuminated and in a gentle breeze or in the dark.
5   Record changes in mass of the seedling assemblies every 10 minutes for 1 hour. (For the purpose of this experiment we assume that any loss in mass of the assembly is due to water vapour escaping from the leaves.)
6   Using the means of the loss in mass for the three assemblies with rooted seedlings and for the three without roots, calculate the rate of loss in mass for each group. Plot a graph of the results.
7   A variation of this investigation is to set up all the seedlings with roots and record the loss in mass for one hour. Then cut off the roots of three of the seedlings, being careful not to get liquid paraffin on the cut stem. Continue to record changes in mass for another hour. Calculate the results as before, and plot a graph.

Questions

a   *Does the presence of roots significantly affect the rate of water vapour loss?*

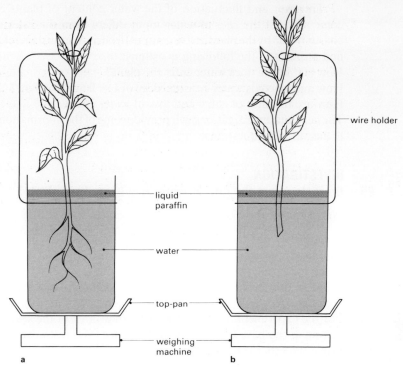

Figure 1
Experimental arrangement for the measurement of water input into seedlings with and without root systems.

b *Under the conditions of your experiment, was the rate of loss in mass constant for the two different assemblies?*

c *If some or all of the leaves were removed, what effect would you predict this would have on the rate of water vapour loss?*

d *How would you establish that the loss in mass was, in fact, due to loss of water in the form of vapour?*

INVESTIGATION
8C Is the rate of water input (and output) affected by environmental conditions around the aerial portions of the plant?

(*Study guide.* Study item 8.11 '*Vegetable staticks*'.)

Figure 2 shows a potometer which can be used to measure the rate of input and output of water into a cut shoot.

Procedure

1. You will need a leafy shoot. Cherry laurel is recommended as it is available all the year round but almost all leafy shoots of woody plants are suitable.
2. Collect shoots the evening before the experiment or early in the morning on the day of the experiment. Select branches of a suitable thickness to fit the rubber bungs, and about 30 cm long. Remove leaves from the lower 10–15 cm of woody stem.
3. During collection, enclose the leafy end of the shoot in a plastic bag and if possible keep the cut end in water.

4. In the laboratory immerse the cut ends of the shoots in a sink full of water. Using a slanting cut remove 3 cm of stem under water, fit the rubber bung, and trim off 3 cm of bark at the cut end of the twig with a scalpel.
5. Leave the prepared stem with its cut end in water until the experimental assembly shown in *figure 2* is prepared.

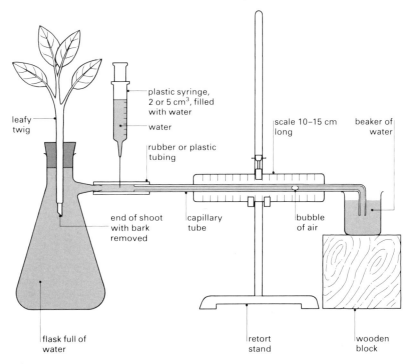

Figure 2
Experimental arrangement for measuring the rate of water input into the cut shoot of a woody plant.

Chapter 8 The plant and water 7

6 To assemble the apparatus, fill the flask, with the capillary tube attached, completely with water. Allow the far end of the capillary tube to dip into a beaker of water.
7 Push the bung with the twig gently into the neck of the flask. This ensures that all air is expelled and excess water forced out via the capillary tube dipping into the beaker of water. The assembly is now ready for experimentation.
8 Introduce a small air bubble into the capillary tube. To do this, lift the free end of the tube out of the beaker of water for a short time, dry it with filter paper, and replace it in the water.
9 Measure the rate of travel of the air bubble under different conditions. For example when the shoot is: illuminated, in the dark, in warm air, covered by a transparent plastic hood, with and without a fan blowing from a distance of 1.5 m. It will take time for the twig to respond to the changed conditions, but as an alternative to waiting for equilibrium rates to be recorded, measured *changes* in rate are equally instructive.
10 When the bubble reaches the end of the scale, reset with the permanently fixed *water filled* plastic syringe as shown in *figure 2*.

Questions

a *How much water passed through the shoot in specified laboratory conditions per unit time?*

b *Under which conditions does water flow vigorously into the shoot?*

INVESTIGATION
8D What is the role of the leaf in the water vapour loss from a cut shoot?

(*Study guide.* Study item 8.11 '*Vegetable staticks*'.)

Procedure A
This is an extension of investigation 8C.
1 Set up the apparatus as in investigation 8C. Arrange it so that the shoot is well lit with or without a gentle fan blowing. (Do not put it under a hood.)
2 Use the apparatus as before and measure the rate of movement of an air bubble in the horizontal capillary tube (*figure 2*). Take several measurements of the time taken for the bubble to move over 5 cm by resetting the bubble with the syringe or by introducing a new one. Rates should not differ by more than 10 per cent from each other and the mean should be calculated.

3 Remove one or two leaves from the shoot using a sharp wet blade. Measure rates of movement of the bubble until a reasonably steady state is obtained. Calculate the mean rate.
4 Remove a further one or two leaves and measure rates of movement of the bubbles as before.
5 Repeat step 4 until all the leaves have been removed.

Procedure B
The role of the leaf in the water vapour loss can also be investigated using a microbalance. This technique assumes that changes in leaf mass are due to water vapour loss by transpiration.

1 Use a microbalance (*figures 3b* and *c* overleaf) to compare the loss in mass of a leaf with that of a piece of wet paper of the same area over the same period. The method for measuring the area of a leaf is shown in *figure 3a* overleaf.
2 Compare the rates of loss in mass of leaves of the same type, but different areas. You can also compare the rates of loss in mass of similar leaves under differing conditions, such as light, dark, under a hood in the light.
3 Smearing petroleum jelly over the surfaces of leaves will produce a waterproof coat. Using this technique investigate whether transpiration occurs mainly through the upper or the lower surfaces of the leaves.
4 Take strips of epidermis, or impressions, from the upper and lower surfaces of the leaves. Peel the epidermis from the leaf under liquid paraffin as this will preserve the state of stomatal opening. Make temporary microscope slide preparations and examine these for stomata. (See *Practical guide 1*, investigation 1A.)

Questions

a *What part do leaves play in the rate of water vapour loss from a shoot?*

b *Does a leaf function as a simple physical evaporator of water, or is there any evidence that the process of water vapour loss from a leaf involves a physiological control mechanism?*

c *Does the rate of transpiration from the surface of a leaf depend on its surface area?*

d *Is the loss of water vapour from a leaf linked with the presence or absence of stomata?*

e *From the results of investigations 8B, 8C, and 8D, summarize what has been established about the input and output of water in plants.*

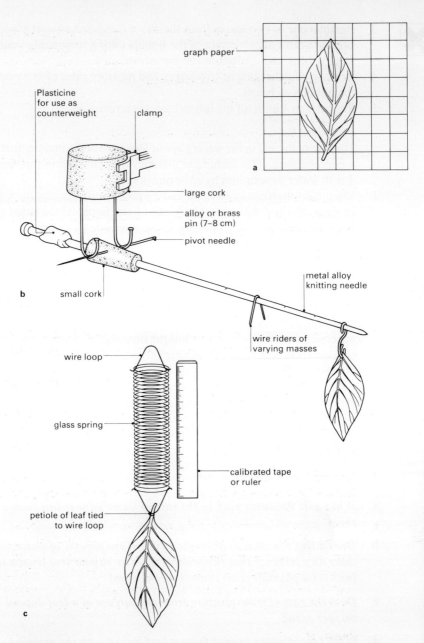

Figure 3
Measuring the area of a leaf (**a**), and recording loss in mass of a leaf using a simple microbalance (**b**) or a glass spring microbalance (**c**).

Having established where the sites of input and output of water are and having obtained an insight into the conditions influencing the rates of input and output, the question of how water is transferred from input to output sites must be answered next.

INVESTIGATION
8E Translocation paths for water in plant stems

(*Study guide* 8.3 'The cohesion theory of water movement in the plant'; Study item 8.31 'The rise of water in a plant'.)

The stem connects the root system of a plant to the leaves and therefore must allow water flowing in from the soil solution to reach the leaves. What tissue in the stem is chiefly concerned with water flow?

Procedure

1. Cut a vigorous shoot, preferably under water or at least with a wet blade, from a plant with a translucent stem, such as 'busy Lizzie' (*Impatiens walleriana, holstii*). It is best to cut the stem when the plant is not transpiring.
2. Set up the stem with its cut end in a 50 cm³ beaker filled with water and support the shoot in an upright position with a retort stand and clamp or stake.
3. Illuminate the shoot for at least 45 minutes with or without a gentle fan blowing from a 1.5 m distance.
4. Use a syringe or bulb pipette to draw out *almost all* the water from the beaker. Be careful to leave 1–2 mm of water at the bottom of the beaker so that the cut end of the stem does not come into contact with air.
5. Replace the water with 10 cm³ of methylene blue dye or eosin solution and continue to expose the shoot to light and air movement for at least an hour.
6. Observe the rise of the dye in the stem (if visible). Cut transverse sections of the stem (*figure 4* overleaf), put them on a microscope slide, and look at them under a microscope. If the sections are too thick, examine them under a dissecting microscope or even with a hand lens in order to identify the nature of the stained cells. It is not necessary to cut complete sections of the stem, small portions are quite adequate.

Questions

a *To which specific tissue of the stem is the dye confined?*

b *Which anatomical structures are found to be stained?*

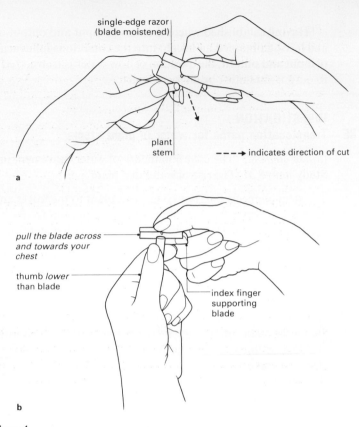

Figure 4
Position of stem and blade when cutting a transverse section through a stem. The stem is held vertically, the razor horizontally.
a View from above.
b View from the front.

c *In your earlier investigations you may have determined some of the conditions under which shoots lose water vapour rapidly. Devise and if possible carry out a further experiment to find out whether the rate at which a dye solution has passed up the stem of a cut shoot is related to the environmental conditions around it.*

INVESTIGATION
8F The anatomy of a stem and how water moves up

(*Study guide* 8.3 'The cohesion theory of water movement in the plant'; Study item 8.31 'The rise of water in a plant'.)

Procedure

1   Take a length of stem and, holding it as shown in *figure 4*, cut thin transverse sections off it using a moistened razor blade. Cut several sections and use the thinnest for further examination. You do not need a complete section across the stem; a small segment will be sufficient. Place the thin section in a little water on a slide.

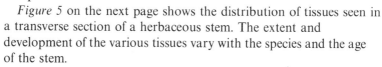

2   The walls of some types of cell in plant tissue become impregnated with lignin. This is a hard substance with a high tensile strength. Stain the preparation for lignin by adding 2–3 drops of benzene-1,3,5-triol (phloroglucinol) to the slide. Leave this for 3–4 minutes, drain off the reagent, and add a drop of concentrated hydrochloric acid. Lignified tissue will be stained red or purple. As other tissues may be distorted by the acid, it is best to make an unstained preparation for comparison.

Figure 5 on the next page shows the distribution of tissues seen in a transverse section of a herbaceous stem. The extent and development of the various tissues vary with the species and the age of the stem.

3   Obtain a second portion of stem which has been left in macerating fluid to soften it (*TAKE CARE*: this is corrosive), wash it thoroughly in water, and place it in a little water on a microscope slide. Using mounted needles, tease the tissues apart. Stain for lignin as before and examine the macerated pieces under a microscope.

Questions

a   *What tissue areas possess cells with lignified walls?*

b   *How is the xylem suited to its function of water translocation?*

Water will rise in fine tubes against the force of gravity. This rise is called capillarity and is due to the physical nature of the liquid surface. An approximate formula to calculate the rise due to capillarity is given by:

$$h = \frac{2\sigma}{rgd}$$

where

h = height to which water will rise in a tube in centimetres
σ = surface tension of water (approximately $10^{-3} \times 73 \, \text{N m}^{-1}$)
r = radius of the tube in centimetres
d = relative density of water (1.0)
g = acceleration due to gravity (approximately $1000 \, \text{cm s}^{-2}$)

4   Find a xylem conducting unit, and, using a micrometer eyepiece and

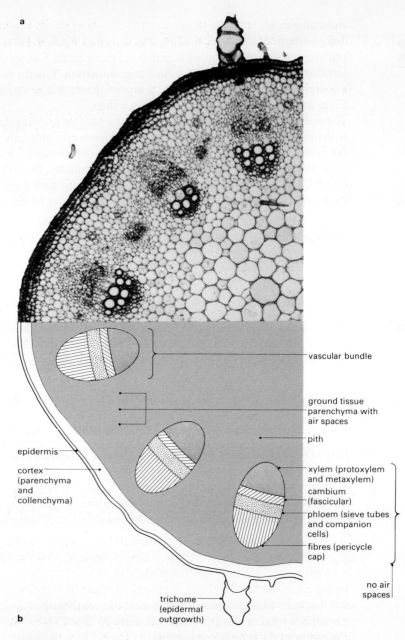

Figure 5
a Transverse section of part of a young dicotyledonous stem (*Helianthus* sp.; ×50)
b Plan of a transverse section of part of a young dicotyledonous stem.
Photograph: **a** *Biophoto Associates.*

14 Cells, tissues, and organisms in relation to water

stage micrometer, measure its average radius. Calculate the height to which water would rise by capillarity, assuming a tube of uniform bore.

c *Is it sufficient to explain how water can reach to the top of the plant?*

d *Do the results of any of your earlier investigations involving the effect of roots and leaves on water movement through the plant help to explain how it may occur?*

INVESTIGATION
8G The xylem pathway

Procedure

1 You need two or three broad bean plants grown to the stage where they have at least three fully expanded leaves.

2 Cut off the stems of the broad beans at ground level, making the cut under water, and keep the cut ends of the stems in water.

3 Fill two or three test-tubes one-third full with a mixed dye solution (equal parts 0.05 per cent eosin and 0.05 per cent Rose Bengal in tap water).

4 Put one shoot in each test-tube and leave in an airy, illuminated place to allow the dye solution to flow into the stem. (A fan heater at a distance of 2 m can be used but do not let the plants get too hot.)

5 After 20–30 minutes remove one shoot from the dye. Beginning on one of the corners of the four-angled stem which bears a leaf, and starting about 2 cm below the leaf, scrape with the blunt edge of a scalpel blade (or single edged razor blade). If the shoots have been in the dye for enough time a thin red line will become visible. This red line represents the xylem part of a vascular bundle.

6 Continue to scrape upwards towards the leaf following the red line. Take increasing care as the node (the point at which a leaf is attached) is approached.

7 The object is to discover where the xylem of that vascular bundle goes. Using a 3-D box (*figure 6* overleaf) as a 'skeleton' sketch the path of the xylem as the dissection proceeds.

8 Having established the pathway of one of the vascular bundles turn the stem through 90° and repeat the process at the next corner trying, by scraping and observation, to follow the pathway of the xylem. Sketch what you see on the 3-D diagram. Repeat the 90° turn twice more, scrape and sketch.

By this stage two 'puzzles' appear:

1 If a bundle disappears into a leaf after bridging the adjacent corner

Chapter 8 The plant and water 15

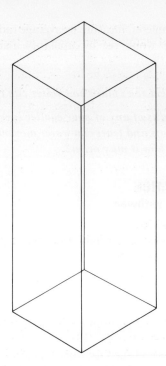

Figure 6
A 3-D box 'skeleton' for drawing the pathway of the xylem along a stem.

bundles where does the corresponding bundle for the next leaf up on the same corner come from? This is best studied by exposing the corner bundle above a leaf and working downwards to see where it comes from (a small amount of cutting may be necessary to remove overlying bundles).

2 Apart from the corner bundles there are several smaller ones (a section taken at an inter-node shows these). Where do these smaller bundles go? Which organs do they serve? Do they join up with each other and with the major corner bundles? These questions are best answered by continuing dissection of the same shoot or by taking a fresh bean shoot and cutting away the corner bundles, allowing the dye to enter only the minor ones. Sketch what you find.

Questions

a *Interpret the final result of your observations and your sketches in terms of*
   1 the supply of water to parts of plants, and
   2 how water supply keeps the shoot erect.

The observations you made in investigations 8F and 8G identified the paths of water movement or translocation. Now we need to explore the forces, or the sources of energy, which cause water to enter into, to move through, and to be lost from plants and cut twigs; work must be done to bring about these changes. We shall investigate whether water input, 'throughput', and output are 'vital' functions (peculiar to living plants) or are based on physical processes.

In principle water movement occurs from regions of relatively high water potential to regions of relatively low water potential, for example water will move from a region where $\psi_s = -0.1\,\text{MPa}$ to one where $\psi_s = -0.2\,\text{MPa}$.

INVESTIGATION
8Ha Water movement by imbibition into a non-living physical system and into live seeds

(*Study guide* 8.2 'Imbibition and osmosis in seeds and seedlings'.)

Procedure

1 Using a cork borer cut at least 10 discs, about 1 cm in diameter, from the plate of gelatine gel provided.
2 Leave the discs on a tile to dry at room temperature for about one day (the time may have to be varied depending on the relative humidity and temperature of the laboratory). A crystal of thymol placed among the discs will prevent fungal growth. (This may have been done for you.)
3 Using blunt forceps take 10 of the partially dried discs and find their mass.
4 Place the discs on wet filter paper in a dish enclosed with cling film. The filter paper must be kept moist by adding water if necessary.
5 Weigh the discs at regular intervals, for example every 30 minutes, and then again on the following day. If the filter paper is not too wet there will be no need to dry the discs before weighing.
6 Repeat the procedure with 10 weighed grains of barley, oats, or wheat, or with pea seeds.
7 Record the results in tabular form and present them as a graph.

Question

a *How could water be extracted from a system (gelatine gel or seeds) which has imbibed water?*

The following procedure is a further investigation into the imbibition of water by seeds.

INVESTIGATION
8Hb Increase in the mass of seeds in different soils and known solutions

(*Study guide* 8.2 'Imbibition and osmosis in seeds and seedlings'.)

Procedure

1. Weigh out seven sets of 20 pea seeds. Keep a record of the dry mass of each set.
2. Put four sets of the seeds into separate shallow dishes and add solutions as follows:

	ψ_s (MPa)
1. distilled water	0.00
2. 1.0 mol dm^{-3} mannitol solution	−2.26
3. 2.0 mol dm^{-3} sodium chloride solution (80% dissociated)	−7.20
4. saturated sodium chloride solution	−30.0

   The seeds must have access to air so do not 'drown' them in the solutions.
3. Put the other three sets of 20 peas into dishes containing the following soils which have been moistened to field capacity (watered thoroughly and left to drain for 48 hours):
   5. clay
   6. sand
   7. garden loam
4. Place all the containers with their seeds in plastic bags to avoid evaporation.
5. After a convenient time (12–48 hours), and before the seeds show any visible signs of germination, remove all seven sets of seeds, blot them with filter paper, and measure and record their masses. This will indicate that the *rate* of water entry is a function of the difference in water potential of the seeds (ψ_{seed}) and that of the external solution ($\psi_{solution}$).
6. Express the gain in mass for each set as a percentage of the gain of mass of the seeds in water.
7. Plot the results as graphs.

Questions

a *What are the water potentials of the seeds, as judged by the inflow of water from solutions with known water potentials?*

b *How do the water potentials of the soils at field capacity compare with the water potentials of the seeds?*

INVESTIGATION
8I A demonstration of the pressure produced during imbibition

(*Study guide* 8.2 'Imbibition and osmosis in seeds and seedlings'.)

Procedure
1 You will be provided with three blocks of gelatine gel which have been partially dried for one or two days at room temperature.
2 Measure each block of gelatine gel exactly and record the measurements.
3 Put each block in a dry shallow dish as shown in *figure 7*. Arrange for a known metal mass, the same in each case, to rest on each block of gelatine. The thread going to the lever must be taut; this can be achieved by weighting the long arm of the lever if necessary.

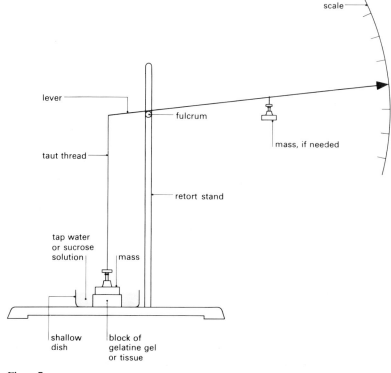

Figure 7
Apparatus set up to demonstrate that pressure (force per unit area) is produced during imbibition.

Chapter 8 The plant and water

4   To one dish add enough tap water to cover the gelatine (not the metal mass); to another dish add 0.6 mol dm^{-3} sucrose solution ($\psi_s = -1.7$ MPa); and to the third dish add 1.0 mol dm^{-3} sucrose solution ($\psi_s = -3.4$ MPa).
5   Measure the changes in the thickness of the gelatine blocks with time and record the position of the lever. Plot the results on a graph.

Question

a   *How does the water potential of the surrounding liquid relate to the pressure produced during imbibition?*

INVESTIGATION
8J Osmotic water movement into a non-living system and into living plant tissue

(*Study guide* 8.2 'Imbibition and osmosis in seeds and seedlings'.)

Procedure A: A non-living system
1   Use a thistle funnel osmometer as shown in *figure 8*. Use 1.0 mol dm^{-3} sucrose solution ($\psi_s = -3.4$ MPa) inside the osmometer and distilled water outside.
2   Measure the rates of meniscus movement into the horizontal capillary tube in cm per 15 or 30 minutes.

Procedure B: Living plant tissue
1   Cut 15 blocks of potato tissue 2 cm × 2 cm × 1 cm and allow them to dry for half an hour to an hour at room temperature. This may have been done for you.
2   Divide the blocks of tissue into three groups of five, blot them gently, weigh and record their mass.
3   Place one set of five blocks in a dish containing tap water, another in 0.1 mol dm^{-3} sucrose solution ($\psi_s = -0.26$ MPa), and the other in 0.25 mol dm^{-3} sucrose solution ($\psi_s = -0.67$ MPa).
4   After 30 minutes remove the blocks from their solutions, blot them gently, weigh and record their mass. Return them to the appropriate solution and continue to determine the change in mass in the same manner every 30 minutes for two hours.
5   Tabulate the results and draw graphs.

Questions

a   *From the results of these experiments give a definition of osmosis in terms of water potential.*

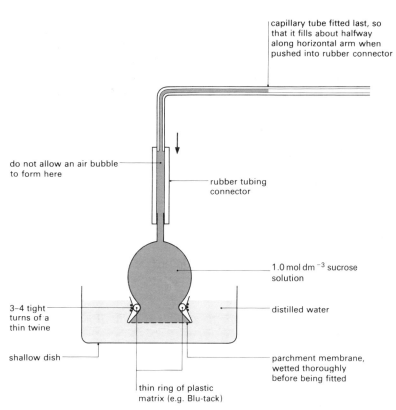

Figure 8
A home-made thistle osmometer. This apparatus is best filled with a syringe.

b *How is osmosis similar to and different from imbibition?*

c *Is there any clear relationship between changes in the mass of the blocks of potato tissue and the water potential of the liquids bathing them?*

d *Instead of finding the mass of the blocks of potato tissue, what other simple measurement could you make which would also indicate changes in their water content?*

e *Using suitable pieces of homogeneous, non-woody plant tissue (like potato tissue), devise an experiment to determine if there is a relationship between the strength of a piece of tissue and its water content. Of what relevance would this be in the living plant?*

INVESTIGATION

8K A demonstration of the osmotic input into seedlings, producing root pressure so that guttation droplets are extruded at special points on the leaves

(*Study guide* 8.2 'Imbibition and osmosis in seeds and seedlings'.)

Procedure

1. You need 10–15 oat, barley, or wheat seedlings, 8–10 cm high, which have been grown in sand in natural daylight and kept well watered.
2. Place a large beaker over the seedlings in the evening and leave at a reasonably constant room temperature.
3. Look at them the next day to see guttation droplets which will have formed at the tips of the leaf blades; some droplets may have run down the blades. If water has condensed on the sides of the beaker remove it very carefully without touching the water droplets. The droplets gradually disappear by evaporation once the beaker is removed.

Question

a. *Guttation occurs more frequently at night. What explanation can you give for this?*

INVESTIGATION

8L Water movement against gravity in a physical system and in plants

(*Study guide* 8.3 'The cohesion theory of water movement in the plant'.)

Procedure

1. The apparatus shown in *figure 9* is called an atmometer. It is used to measure the evaporating power of the air. When placed on a top-loading weighing machine, evaporative loss from the porous pot can be measured by monitoring changes in mass. Alternatively, the atmometer can be left on the bench and the volume of water lost from the reservoir determined by filling it up to its original level with water delivered from a graduated syringe or pipette.
2. Use an atmometer to make measurements of the evaporative power of the air under the following conditions:
   1. light and dark (if using an artificial light take care not to heat the porous pot; the light source should be at least 40 cm away)
   2. still air and moving air
   3. porous pot covered by a plastic hood.

Böhm's experiment described in *Study guide I*, section 8.3, is a

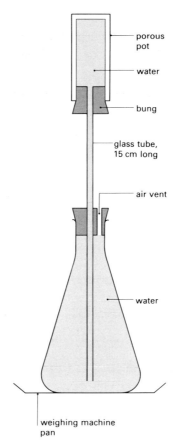

Figure 9
An atmometer used for measuring the evaporating power of the air.

variation on an atmometer in which the vertical rise of water in a capillary tube is the object of measurement. *Figure 10* overleaf shows simplified versions of Böhm's apparatus.

3   Use this apparatus (a) with a porous pot and (b) with a leafy shoot to measure the rise of the dye in the capillary tube under the same conditions as those used with the atmometer in step **2** above. Allow time for adjustment to each set of conditions before taking a reading.

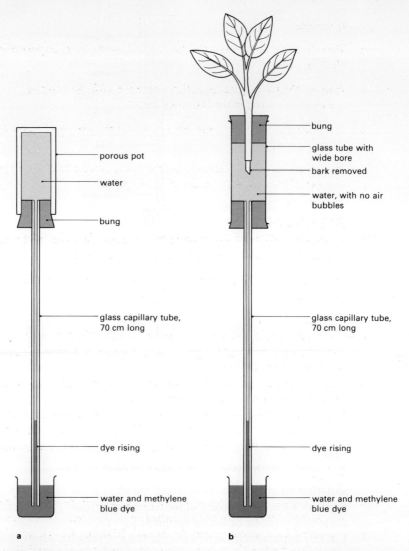

Figure 10
Simplified versions of Böhm's apparatus for demonstrating the ascent of a column of water, promoted by
a vapour loss from a porous pot, and
b 'transpiration pull' due to vapour loss from a leafy shoot.

Questions

a *How do environmental conditions affect the rate of evaporation from the atmometer?*

b *How do the rates of evaporation compare with the rates of transpiration under similar conditions?*

c *What do these experiments tell you about the ascent of water as a phenomenon?*

d *What information do these experiments provide about the degree of control of transpiration exerted by the plant?*

INVESTIGATION

8M A demonstration of the mechanical support function of turgor pressure produced by water input into leaf cells

(*Study guide* 8.4 'What role does water play in the life of a plant?'.)
This is best demonstrated with a lupin leaf.

Procedure

1   Cut a good lupin leaf with a petiole about 20–30 cm in length.
2   Cut 2 cm off the end of the petiole under water and keep it immersed in water for about 10 minutes.
3   Remove the petiole from the water and allow the leaf to wilt visibly under illumination and in a gentle air current.
4   Replace the petiole in the water. Cut 2–3 cm off the petiole under water and observe the recovery of the leaf as it regains turgor. The movement of the leaflets can be seen with the naked eye within two minutes.

Question

a *Large-leaved plants such as sugar beet often wilt temporarily in temperate climates — this is called 'mid-day wilt'. It cannot be remedied by applying water to the soil. What is the explanation for this wilting when there is no shortage of water in the soil? What could be done to prevent it happening?*

CHAPTER 9 **THE CELL AND WATER**

INVESTIGATION
9A Measurement of the water potential of cell sap by plasmolysis

(*Study guide* 9.1 'Water relations of a plant cell'; Study item 9.11 'Quantitative examples of water potential in plant cells'.)

Procedure

1. Prepare a series of sucrose solutions of different concentrations from $0.2\,\text{mol}\,\text{dm}^{-3}$ to $1.0\,\text{mol}\,\text{dm}^{-3}$. The most convenient way of doing this is to put distilled water in one burette and $1.0\,\text{mol}\,\text{dm}^{-3}$ sucrose in another and make up $0.2\,\text{mol}\,\text{dm}^{-3}$, $0.4\,\text{mol}\,\text{dm}^{-3}$, $0.6\,\text{mol}\,\text{dm}^{-3}$, and $0.8\,\text{mol}\,\text{dm}^{-3}$ by running appropriate quantities into specimen tubes. Then put the prepared solutions into watch-glasses.

2. Prepare, either thin hand sections of beetroot (50 mm thick) or pieces of epidermis (0.5 cm × 0.5 cm) containing anthocyanin (from rhubarb, pink onion, or cyclamen). Place three pieces in each solution. Swirl them about to make sure the sections or epidermal strips are immersed. Leave for 5–10 minutes.
3. After this interval remove the pieces from the solution one at a time. Mount each piece on a microscope slide in a few drops of the solution in which it was immersed and add a coverslip. Examine the tissue under the high power objective.
4. Count the total number of cells visible in the field, and then count those which are plasmolysed. These are cells in which the cytoplasm has come away from the cell wall as shown in *figure 11*.
5. Plot a graph of percentage of cells plasmolysed against molarity of sucrose.
6. It is very instructive to place some of the plasmolysed tissue into a drop of tap water and observe under the microscope.

Questions

a *From your graph, read off the molarity of sucrose which corresponds to 50% plasmolysis. What does this represent?*

Table 1 gives the water potential in MPa of sucrose solutions.

b *What was the water potential of the cell sap at incipient plasmolysis in the tissue you used?*

c *When 50% of the cells in a piece of tissue are plasmolysed what is the average turgor pressure within the cells?*

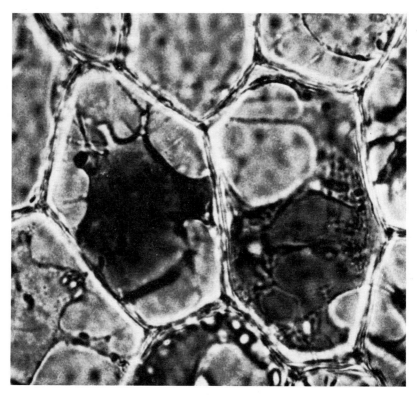

Figure 11
Plasmolysed cells (× 1500).
Photograph, Professor Hans Meidner.

Concentration (mol dm^{-3})	Water potential (ψ_s) (MPa)
0.1	−0.26
0.2	−0.54
0.3	−0.82
0.4	−1.12
0.5	−1.45
0.6	−1.70
0.7	−2.17
0.8	−2.58
0.9	−3.01
1.0	−3.40

Table 1
Water potential of solutions (ψ_s) of given molar concentrations of sucrose at 20 °C.

d *What inaccuracies are inherent in this method of determining water potential of plant cells?*

e *What happened to the plasmolysed cells when they were placed in tap water and what does this demonstrate about the state of the cells?*

INVESTIGATION
9B Diffusion of substances against gravity in a gel

(*Study guide* 9.1 'Water relations of a plant cell' – see 'Diffusion', page 270.)

Procedure

1 Take two test-tubes filled to the brim with 3 per cent gelatine. Fix a cm scale to each tube.
2 Using a test-tube rack for support, invert one of the tubes with its mouth over crystals of copper(II) sulphate and the other over a few drops of eosin or methylene blue in a small Petri dish.
3 Take a third tube filled with 3 per cent gelatine to which Universal Indicator has been added (it will be pale green in colour). Add a scale and invert this over a few drops of $0.5\ mol\ dm^{-3}$ hydrochloric acid in a small Petri dish.
4 Measure the height the colour reaches in the tubes in each case after known times up to 24 hours.

Questions

a *Explain what appears to have happened in each case in terms of movement of particles. Which particle moves fastest and why?*

b *At what point would an equilibrium be established?*

INVESTIGATION
9C Movement through a differentially permeable membrane

(*Study guide* 8.2 'Imbibition and osmosis in seeds and seedlings'; 9.1 'Water relations of a plant cell' – see 'Dialysis' page 271; 10.6 'Kidney failure'.)

Procedure

1 Take two Visking tubing bags, one containing pale iodine solution and suspended in a measuring cylinder also containing pale iodine solution, and the other containing 0.2 per cent soluble starch solution suspended similarly in 0.2 per cent starch solution.

2   Lift out the bag containing iodine solution; hold it with a bulldog clip and do not squeeze it. Wash the bag very thoroughly under running tap water. Do the same to the bag containing starch solution.
3   Place the bag containing the iodine solution in the measuring cylinder of starch solution and put the bag containing starch solution into the iodine solution. Hold the bags with bulldog clips as before and avoid spilling the contents by squeezing the bags; keep the mouths of the bags clear of the liquids in the measuring cylinders.
4   Observe the coloration inside and outside the bags during the next 5, 10, and 15–30 minutes.

Questions

a   *What appears to have happened in terms of movement of particles? Which kind of particle penetrates the membrane?*

b   *How is this process similar to and different from the diffusion which occurred in the previous experiment?*

c   *In what way is this process, which is called dialysis, similar to and different from osmosis, which was demonstrated in investigation 8J?*

Dialysis has important industrial and medical applications (see account of artificial kidney in *Study guide I* page 310).

INVESTIGATION
9D The effect, on isolated animal cells, of altering the composition of the external medium

(*Study guide* 9.2 'Water relations of animal cells; Study item 9.21 'The effect of various solutions on human red blood cells'.)

You can test a theoretical model of the water relationships of an animal cell experimentally by exposing isolated animal cells to different concentrations of water in their external environment. Red blood cells are convenient for this purpose.

Procedure

1   Dilute the blood provided 1:10 with iso-osmotic saline. Put 1 cm^3 of the diluted blood into each of three clean dry test-tubes labelled A, B, and C.
2   Add 10 cm^3 of the distilled water to tube A, 10 cm^3 of iso-osmotic saline to tube B, and 10 cm^3 of 1.0 mol dm^{-3} saline to tube C.

3   Leave for a few minutes, and then compare the turbidity of the solutions. You can do this either by using a colorimeter or by placing a sheet of small typescript behind the tubes and seeing how easily it can be read.
4   Take a drop of liquid from tube A, put it on a microscope slide and add a clean coverslip. Examine it under a microscope. Repeat this with the other two tubes.

Questions

a   *What was the effect of each of the reagents on the red blood cells? Relate the turbidity of the liquids to the appearance of the cells as seen under the microscope.*

b   *Explain the observed effects in terms of a hypothetical model of an animal cell as given in the* **Study guide** *(figure 192, page 272).*

c   *What does this investigation suggest about the composition of the blood plasma (the natural external medium of red blood cells)?*

d   *In what ways do animal cells behave differently from plant cells, when immersed in solutions of different concentrations? What is the reason for the difference?*

INVESTIGATION
9E   The action of a contractile vacuole

(*Study guide* 9.2 'Water relations of animal cells'; Study item 9.22 'The action of the contractile vacuole complex'.)

The contribution that contractile vacuole complexes make to the osmoregulatory activities of single-celled fresh water organisms can be illustrated by observing the response of these organisms to various solutions of different concentrations. A series of experiments can be carried out using either the suctorian *Discophrya* (*figure 12*) or the more familiar *Paramecium* (see the cover photograph).

Procedure

1   Make up a series of solutions of sucrose ranging from 0 to 0.1 mol dm^{-3} by running appropriate quantities of distilled water and 0.1 mol dm^{-3} sucrose from the burettes into small dishes.
2   To each dish add either several silk threads or hairs bearing *Discophrya* (*figure 12*) or an equal volume of an actively growing *Paramecium* culture.
    (*Note.* In the case of *Paramecium* the range of concentrations of the experimental solutions will then be from 0 to 0.05 mol dm^{-3}.)

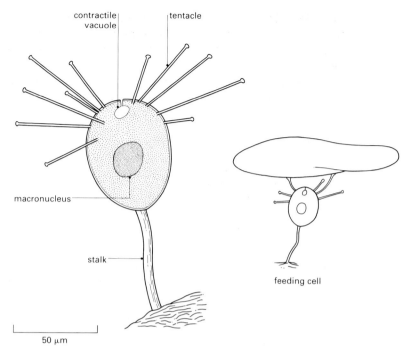

Figure 12
Discophrya attaches to the substrate by an inert stalk and feeds by means of numerous tentacles. There is a large macronucleus in the centre of the cell, and the single contractile vacuole is positioned near the unattached apex of the cell.
Based on a drawing by Dr David J. Patterson, Department of Zoology, University of Bristol.

3   After a few minutes remove cells from one of the more concentrated experimental solutions (*e.g.* 0.06 mol dm^{-3} for *Discophrya* and 0.05 mol dm^{-3} for *Paramecium*). Observe and draw their appearance and behaviour. Compare them with cells from the normal culture medium.

a   *What has happened to these cells?*

4   Leave the cells in their experimental solutions for one hour to equilibrate.

5   While waiting, place a drop of *Paramecium* culture on a slide and add a coverslip. Gently remove some fluid from between slide and coverslip with the corner of a paper tissue. This will draw the coverslip towards the slide and will gently press on the cells to stop them moving. They can then be observed easily. If the cells are still mobile, remove some more fluid; if they have broken apart, begin again. Watch the

behaviour of one of the contractile vacuoles, and compare it with *figure 13*. Sustained pressure will block the pore of the vacuole and after a while many contractile vacuoles will become abnormally distended.

b   *Do both contractile vacuoles behave similarly?*
    Is there any evidence that these vacuoles really are contractile?

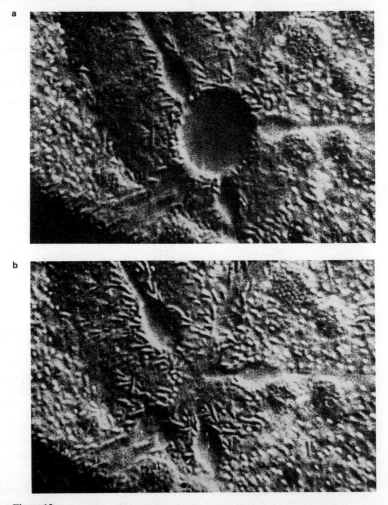

Figure 13
The contractile vacuole of *Paramecium* photographed (**a**) at the end of the filling part of the cycle (diastole) and (**b**) at the end of the phase of expulsion (systole). The radiating collecting canals can be seen easily, as can the distensible ampullae. When the vacuole collapses it can no longer be seen when viewed from above, as here. (× 1800.)
Photographs, Dr David J. Patterson, Department of Zoology, University of Bristol.

6 After allowing for experimental cells to equilibrate for one hour (this may have been done for you) you can monitor the activity of the contractile vacuoles. Remove some cells in a small drop of water and put them on a slide. (To impede the movement of *Paramecium* add either a drop of 2.5 per cent methyl cellulose, or some teased out lens tissue or cottonwool.) Record the activity of one contractile vacuole by measuring the duration of the cycle, or by measuring the total output. To measure the output you will need to measure the size of the vacuole just before it expels its contents.

7 Repeat with other experimental solutions. Express your results graphically, and compare them with *figure 14*.

c *Does measuring the frequency of contractions of the contractile vacuole provide a satisfactory measure of the osmoregulatory activity of this organelle?*

d *Why is the relationship between vacuolar activity and external water potential not linear?*

8 Look again at cells from the medium used in step **3**.

e *Draw some typical cells. Has their appearance changed in any way? Why?*

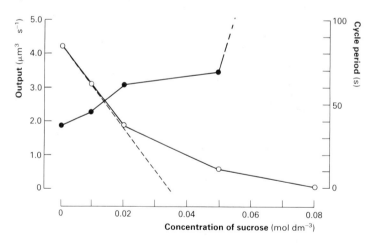

Figure 14
The behaviour of the contractile vacuole complex of *Discophrya* is here shown as a function of the water potential of the bathing medium. As the water potential of the medium is decreased so the overall output of the complex (open circles) drops and the duration of the cycle (closed circles) gets longer. The output of the vacuole which would be expected if the cell behaved as a perfect osmometer is shown by the dashed line.
Based on a drawing by Dr David J. Patterson, Department of Zoology, University of Bristol.

CHAPTER 10 CONTROL BY THE ORGANISM

INVESTIGATION
10A The relation of the urinary system of a mammal to other systems of the body

(*Study guide* 10.3 'The internal environment'.)

The structural relationship of the urinary system to the rest of the body is very similar in mammals, and therefore any convenient mammal such as rat or mouse is suitable for this investigation. The following instructions are somewhat generalized and should cover all mammals likely to be used.

For reasons of economy the reproductive system should be investigated at the same time as the urinary system, and in order to use as few animals as possible half the class should dissect a male and half a female. Make sure that you have the opportunity of studying the reproductive organs of both sexes. (See *Study guide II*, Chapter 22.)

Procedure

1. Pin the dead animal down, through the limbs, with the ventral side uppermost. Identify the urinary and genital openings in the male and female and note any other external differences between the sexes.
2. Carefully open the animal with a ventral abdominal incision, cutting through the skin and body wall to expose the alimentary canal as shown in *figure 15*.
3. Ligature the hepatic portal vein and cut through the oesophagus and rectum (*Practical guide* 1, investigation 1E). Gently remove the whole gut and liver from the abdominal cavity by cutting the mesenteries and blood vessels. Any bleeding can be stopped by the application of a little alcohol on cottonwool.
4. Locate the kidneys, examine the blood supply to and from them, and identify the renal arteries and veins. Trace the ureters to the bladder. It may be necessary to remove a certain amount of fat. Note the position of the adrenal glands (see *figures 18* and *19*).

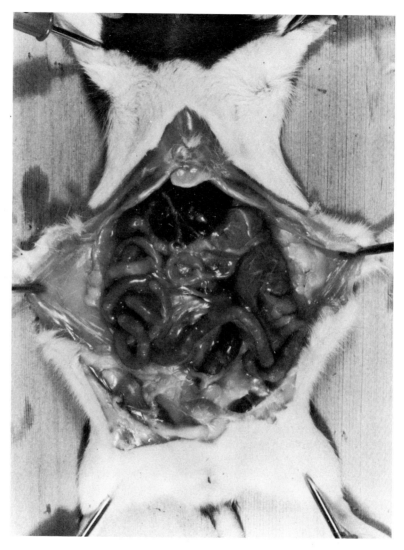

Figure 15
Dissection of a rat, with the alimentary canal exposed.

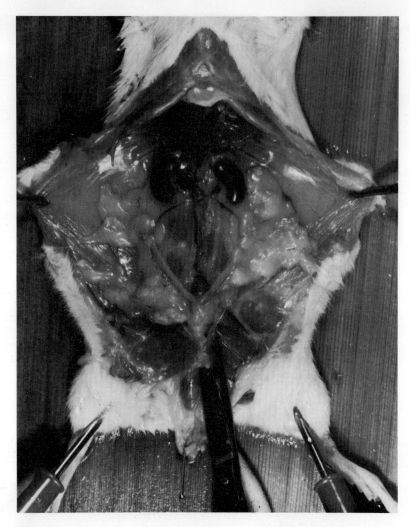

Figure 16
Dissection of a rat. The scissors are in position to cut one side of the pelvic girdle.

5   To trace the urinary system to its external opening you need to remove part of the pelvic girdle. To do this, find the position of the pelvic girdle and then, holding a pair of scissors almost horizontal cut through it a little to the right of centre as shown in *figure 16*. Then make a similar cut on the left side and remove the central portion. Pin out the knees to clearly expose the region, and then trace the urethra to its external opening.

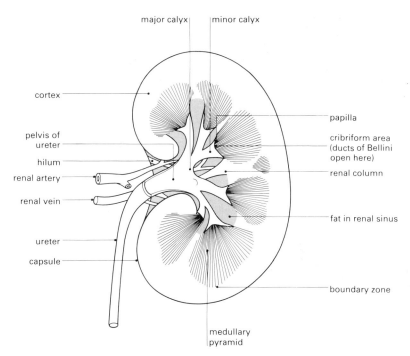

Figure 17
The structure of the human kidney.

Macrostructure of the kidney

You can remove a kidney and the upper part of its ureter from the mammal you are dissecting, but it is better to use a pig's or lamb's kidney from a butcher as it is larger.

6   Slice through the kidney horizontally, that is, parallel to its dorsal surface, so as to cut across the portion where the ureter leaves it. Examine the cut surface with a hand lens.

7   Identify the parts of the kidney with the help of *figure 17*, and then make a labelled drawing of your own specimen.

Questions on the urinary system.

a   *From what part of the kidney does the ureter leave and where does it join on to the bladder?*

b   *To what is the artery supplying the kidney with blood connected, and to what is the vein carrying the blood from the kidney connected?*

c   *From the relationships of the arterial blood supply, what hypothesis can you make about the pressure of blood entering the kidney?*

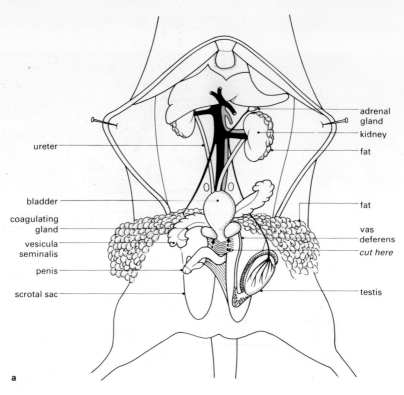

Figure 18
Reproductive organs of a male rat: **a** in relation to other parts of the body; **b** (*opposite*) showing more detail.
Based on Rowett, H. G. Q., Dissection guides III: The rat, with notes on the mouse, 2nd edn, John Murray, 1952.

Procedure for the male reproductive system

8 Open a scrotal sac and note the testis anchored in position by the gubernaculum (a band of tissue by whose contraction the testis is withdrawn from the dorsal side of the abdomen into the scrotum). Find the epididymus, vas deferens (sperm duct), seminal vesicle, prostate gland, and penis. (See *figure 18*.)

9 Make a labelled drawing of your dissection of the male urinogenital system.

Questions on the male reproductive system

a *What is the relationship between the ureters (which transmit urine from the kidneys), the vasa deferentia (which transmit sperms from the testes), the urethra, and the penis?*

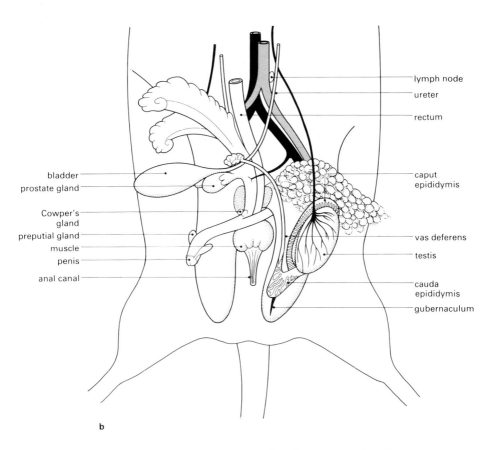

b

- **b** *What functions can you suggest for the prostate gland?*
- **c** *Is there any evidence to show that the testes were originally located in the abdominal cavity but, as the animal developed, descended into the scrotal sacs?*
- **d** *What indication, if any, did you find of a blindly-ending sac (resembling a uterus) extending forwards into the abdominal cavity and ending at a point where the vasa deferentia and urethra join?*
- **e** *A rudimentary uterus (known as the uterus masculinus) is sometimes well developed in certain male mammals, notably the guinea pig. What conclusion can you draw from the existence of an element of femaleness in a normal male animal?*

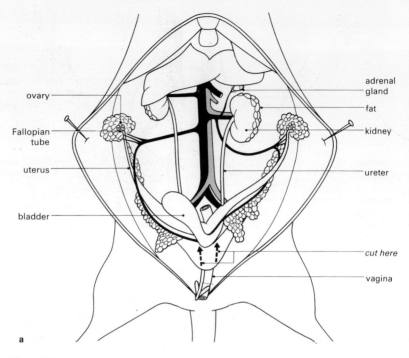

Figure 19
Reproductive organs of a female rat: **a** in relation to other parts of the body;
b (*opposite*) showing more detail.
Based on Rowett, H. G. Q., Dissection guides III: The rat, with notes on the mouse, 2nd edn, John Murray, 1952.

Procedure for the female reproductive system
8 Find the ovaries, Fallopian tubes, uterus, and vagina (*figure 19*).
9 Make a labelled drawing of your dissection of the female urinogenital system.
10 Cut through the uterus and a Fallopian tube and compare the thickness of their walls.

Questions on the female reproductive system

a *In the light of your findings, where do you think it is most likely that fertilization and the development of the embryo take place?*

b *How does the function of the urethra in the female differ from that in the male?*

c *What indication, if any, is there of a male portion among the female reproductive organs? How do you reconcile your findings with question d above?*

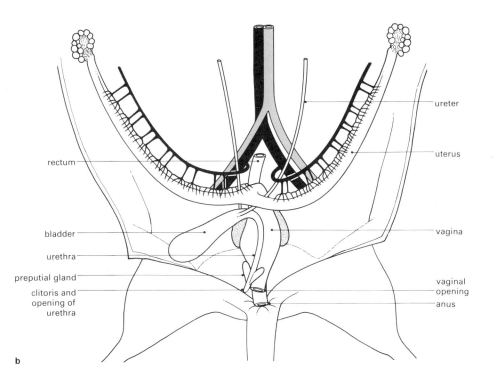

b

INVESTIGATION
10B Injection of the arterial blood system in the kidney

(*Study guide* 10.4 'The functioning of the kidney'.)

Gloves and goggles should be worn during this investigation.

Kidneys can be injected with colouring matter through the main blood vessels to provide a more or less permanent record of the internal blood supply (*figure 20*).

Procedure

1. For this investigation you require lamb's kidneys fresh from a butcher. It is best to have kidneys still enclosed in fat. Remove the fat carefully and locate the renal artery, renal veins, and ureter.
2. The simplest procedure is to inject coloured latex into the renal artery. First, you must inject 3 to 5 cm^3 of warm Ringer's solution. A convenient way of doing this is to use a 5 cm^3 hypodermic syringe fitted with a short length of thin rubber tubing. Attach this to a short piece of thin glass tubing drawn out at the end so that it is small enough to fit the artery. Tie cotton firmly round the artery and glass tubing. As you inject the Ringer's solution you should feel the kidney becoming distended.

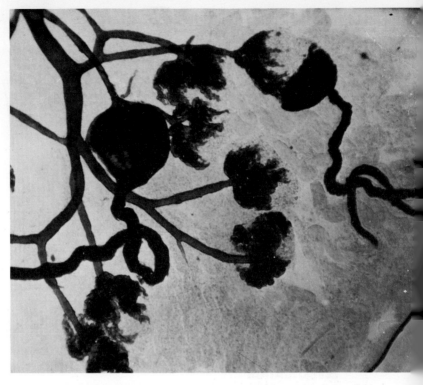

Figure 20
Injected kidney tissue showing afferent arteries, glomeruli, Bowman's capsules, and parts of the renal tubules. (× 200.)
Photograph, J. F. Eggleston and M. Bonsir.

3 Detach the syringe from the rubber tubing and take up a similar quantity of warm red latex into the syringe. Inject this slowly and you should see it under the surface of the kidney as it reaches the glomeruli in the cortex. Remove the glass tube and tie cotton round the renal artery.

4 Place the kidney in 2.0 mol dm^{-3} hydrochloric acid for 24 hours. Wash acid off the outside of kidney before handling it.

5 Slice the kidney in the flat plane, parallel with the exit of the ureter. Examine it under a low power stereoscopic binocular microscope.

6 Select a portion of the cortex showing the injected blood vessels clearly. With a scalpel slice a section about 3 mm thick parallel to the flat surface.

7 Place this section in fresh 2 per cent pepsin in a specimen tube. The hydrochloric acid already present should be sufficient to bring the pH down to about 2 or 3. Close the tube and leave for two to three days.

After this time remove the section and rinse it gently in water. Using a pair of forceps pick out suitable blood vessels and mount them in water on a slide. Examine under a microscope.

Questions

a   *What parts of the kidney structure can you identify? Make a sketch of these.*

Look at a diagram of kidney structure, such as figure 210b of *Study guide I*.

b   *Compare the distribution of the branches of the renal artery, supplying the glomerulus, with a conventional diagram of the structures. In what way does the distribution in your preparation differ from the diagram?*

c   *What can be inferred from the preparation, about the resistance to flow in the glomerulus blood vessels?*

d   *Does the preparation give any indication of the nature of the barrier between the blood vessels of the glomerulus and the lumen of Bowman's capsule?*

Additional investigations

Injection of venous blood system
It is possible to extend the injection technique by using the method described to inject warm blue latex into the renal vein. You can even try injecting both artery and vein in one kidney.

Injection via the ureter has not proved to be very satisfactory as the latex accumulates in the pelvis of the kidney.

A whole kidney preparation
This is a long-term investigation requiring several weeks to complete.

Follow steps **1–4** of the procedure already described injecting either or both of the blood systems.

5   Place the whole kidney with its outer capsule removed in 2 per cent pepsin in a beaker. Leave several days. Remove and wash in running water.

6   Replace the kidney in fresh pepsin and repeat the washing and pepsin treatment until the kidney tissue is sufficiently digested to be removed leaving just the injected latex.

7   Put the whole preparation in clean water in a beaker. Remove pieces of the latex and examine them under a microscope.

This whole preparation gives a very vivid idea of the total blood supply of the kidney. The injection technique is similar to that used by Bowman (1842) although for his injection material he used lead chromate.

INVESTIGATION
10C The histological structure of the nephron

(*Study guide* 10.4 'The functioning of the kidney'.)

As various parts of the nephron perform different functions we might expect the tubule to show a different structure along its length. As the nephron is a three-dimensional structure, it is impossible to see all of the structure from the examination of a single section. You would need to study a series of sections cut consecutively through the kidney and build up a reconstruction of a nephron from a study of these. This is a time-consuming task. However, a careful study of microscope preparations or 35 mm transparencies of them does provide some useful first-hand information on the structure of the nephron.

Procedure

1. *Figure 21* is a diagrammatic reconstruction of a nephron. Use this with your slide or transparency as an aid to identification.
2. Examine a microscope slide preparation (or transparency) of a kidney section under low power magnification. Identify cortex and medulla and note the characteristic appearance of each region.

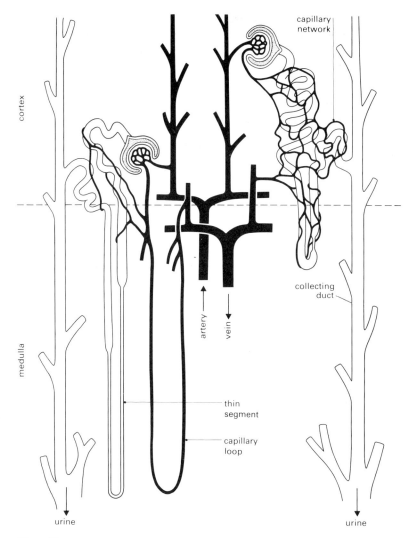

Figure 21
Diagram showing two nephrons. The long-looped nephron is paralleled by a loop formed by the blood capillary. The short-looped nephron is surrounded by a capillary network. Most mammalian kidneys contain a mixture of the two types of nephrons, but some species have only one or the other kind.
Based on Schmidt-Nielsen, K., Animal physiology: adaptation and environment, Cambridge University Press, 1975, as drawn from Plate II of Smith, H., The kidney, Oxford University Press, 1951.

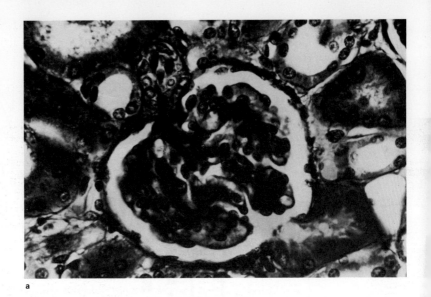

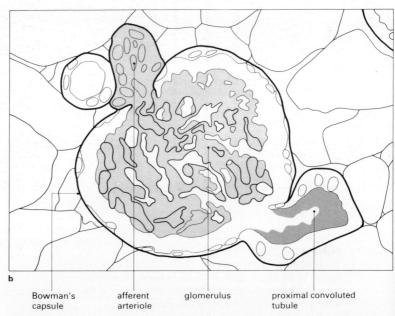

| Bowman's capsule | afferent arteriole | glomerulus | proximal convoluted tubule |

Figure 22
a Photomicrograph of a renal corpuscle (× 460).
b Explanatory drawing of the corpuscle.
From Wheater, P. R., Burkitt, H. G., and Daniels, V. G., Functional histology – a text and colour atlas, Churchill Livingstone, 1979.

3 Then examine the cortex under high power magnification. Locate a glomerulus and refer to *figure 22* to help you with the interpretation of your slide.

Scanning electronmicrographs (*figure 23*) show the three-dimensional nephron as well as details which cannot be seen with a light microscope, such as podocytes (*Study guide* 10.4).

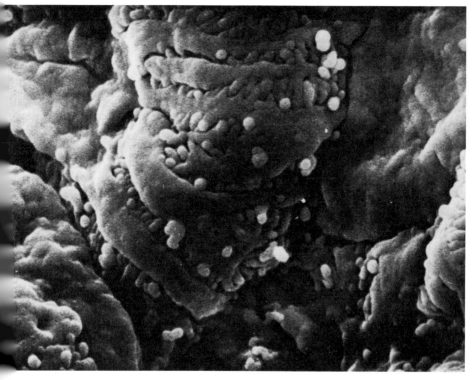

Figure 23
Scanning electronmicrograph of a glomerulus, showing podocytes (× 6000).
From Wheater, P. R., Burkitt, H. G., and Daniels, V. G., Functional histology *– a text and colour atlas,* Churchill Livingstone, 1979.

4 Examine the tubules of the cortex which are a mixture of proximal and distal convoluted tubules and collecting ducts. The proximal convoluted tubule arises from the renal corpuscle. With the help of *figure 24* identify its special characteristics.

Again electron microscopy reveals a great deal more detail, as shown in *figure 25*.

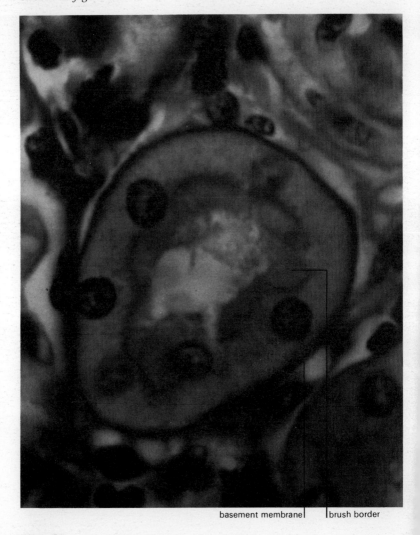

basement membrane | brush border

Figure 24
A transverse section of a proximal convoluted tubule showing the brush border of microvilli ($\times$ 1700).
From Wheater, P. R., Burkitt, H. G., and Daniels, V. G., Functional histology – a text and colour atlas, *Churchill Livingstone, 1979*.

Figure 25
An electronmicrograph of a proximal convoluted tubule of a kidney, showing the microvilli which constitute the brush border (× 6000). *Photograph, J. H. Kugler.*

The distal convoluted tubule differs from the proximal tubule in several ways, one of the most obvious being the absence of a brush border (*figure 26*). The collecting ducts can be distinguished from the proximal and distal convoluted tubules by their well defined cellular outline.

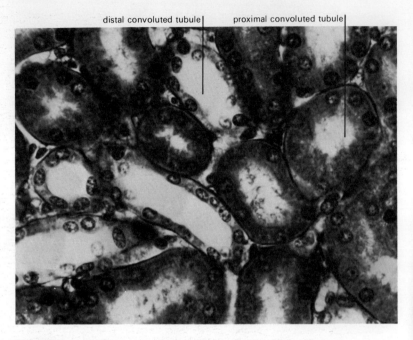

Figure 26
In transverse section the distal convoluted tubule can be distinguished from the proximal convoluted tubule by the absence of a brush border and the presence of a larger lumen. (× 525.)
From Wheater, P. R., Burkitt, H. G., and Daniels, V. G., Functional histology – a text and colour atlas, *Churchill Livingstone, 1979.*

5   Now look at the medulla of the kidney under high power magnification (*figure 27*). This consists of various tubules including the loops of Henle, which are a rounded shape. The thin descending loops have simple flattened epithelial walls and the thick ascending

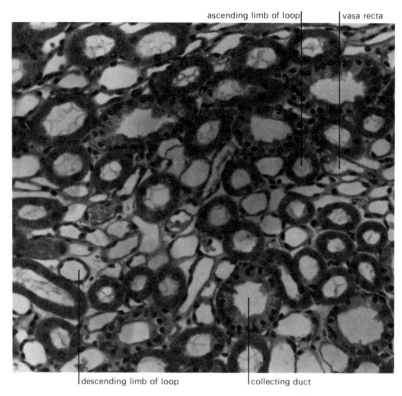

Figure 27
Transverse section through loops of Henle. The thin descending limb is lined with a flattened epithelium. The ascending limb is thicker and has a cuboidal epithelial lining. The vasa recta are thin but contain blood cells. The collecting ducts have a columnar epithelial lining and are wider. (× 250.)
From Wheater, P. R., Burkitt, H. G., and Daniels, V. G., Functional histology – a text and colour atlas, *Churchill Livingstone, 1979.*

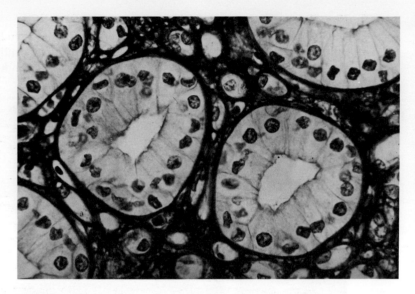

Figure 28
Transverse section of collecting ducts, showing the columnar epithelial lining. (× 465.)
From Wheater, P. R., Burkitt, H. G., and Daniels, V. G., Functional histology – a text and colour atlas, *Churchill Livingstone, 1979.*

loops have walls made up of cuboidal cells. The collecting ducts are distinguishable by their columnar epithelium (*figure 28*).

6   Having identified the structures, draw them and tabulate the appropriate data alongside.

Questions

a   *From the examination of sections of kidney and photomicrographs, what can you deduce about the function of various parts of the nephron?*

b   *How can more precise evidence about the function of the various parts of the nephron be obtained?*

STUDY ITEM

10C1   (Multiple choice exercise)   (J.M.B.)

The following drawings, which are not to scale, represent transverse sections of tubes found in the kidney. Which one of the rows A to D correctly identifies the tubes?

52   Cells, tissues, and organisms in relation to water

	Figure 29	Figure 30	Figure 31

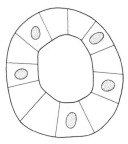

Diameter: 50–60 μm   15–20 μm   50–60 μm

A	distal convoluted tubule	capillary	proximal convoluted tubule
B	collecting duct	base of loop of Henle	distal convoluted tubule
C	proximal convoluted tubule	base of loop of Henle	collecting duct
D	collecting duct	capillary	proximal convoluted tubule

INVESTIGATION

10D Determination of chloride content of urine collected in different circumstances of salt intake

Students must only handle and dispose of their own urine samples.

The kidney performs its osmoregulatory function by selectively reabsorbing water and salts from the kidney tubule back into the blood. The control of this is discussed in *Study guide* 10.5.

With a little preliminary organization you can collect samples of urine after different intakes of salt and compare their chloride content using a titration technique.

Procedure

1. The collection of the urine must be carried out before the practical session. Collect the urine in boiling tubes, with bungs, labelled specimen 1 and specimen 2. A few drops of methylbenzene (toluene), which acts as a preservative, should be added to each tube unless the titration is to be carried out immediately.
2. Urinate about one hour before you have a meal. Eat plenty of salt with your meal but do not drink any liquid.
3. About one hour later collect 5–10 cm³ of your urine as the first specimen.

4 Then drink as much water as possible and after about another hour collect specimen 2.
5 For the titration, put $0.1\,\text{mol dm}^{-3}$ potassium thiocyanate into a burette.
6 Use a syringe to transfer $2\,\text{cm}^3$ of urine to a conical flask. Add $10\,\text{cm}^3$ of $0.1\,\text{mol dm}^{-3}$ silver nitrate solution and stir with a glass rod. Any chloride will be precipitated as white silver chloride.
7 Leave the mixture to stand for about five minutes to allow the precipitate to coagulate and then add a few drops of a saturated solution of iron(III) nitrate which is the indicator.
8 Carefully run $0.1\,\text{mol dm}^{-3}$ potassium thiocyanate from the burette into the conical flask. Stir gently all the time until a red colour appears and remains for 15 seconds. Note the volume of thiocyanate required.
9 Repeat the titration with the second sample of urine.

Calculation of results
The reactions involved in this titration are:
1. Excess silver nitrate is added to a known volume of urine to precipitate the chloride present:
$AgNO_3 + NaCl \rightarrow AgCl + NaNO_3$
2. Potassium thiocyanate reacts with unprecipitated silver nitrate:
$KCNS + AgNO_3 \rightarrow KNO_3 + AgCNS$
3. The first excess drop of thiocyanate reacts with the indicator, iron(III) nitrate, forming red iron(III) thiocyanate:
Iron(III) nitrate + KCNS → iron(III) thiocyanate + KNO_3
The solutions of potassium thiocyanate and silver nitrate used are both $0.1\,\text{mol dm}^{-3}$ and so are equivalent. Therefore, if $x\,\text{cm}^3$ of potassium thiocyanate (KCNS) are required to precipitate the unused silver nitrate ($AgNO_3$) then $(10-x)\,\text{cm}^3$ have been precipitated by the chloride in $2.0\,\text{cm}^3$ of urine.

$1\,\text{dm}^3\ 1.0\,\text{mol dm}^{-3}\ AgNO_3$ is equivalent to $35.5\,\text{g}$ chloride

so $1\,\text{dm}^3\ 0.1\,\text{mol dm}^{-3}\ AgNO_3$ is equivalent to $\dfrac{35.5}{10}\,\text{g}$ chloride, and

$1\,\text{cm}^3\ 0.1\,\text{mol dm}^{-3}\ AgNO_3$ is equivalent to $\dfrac{35.5}{10} \times \dfrac{1}{1000}\,\text{g}$ chloride

so $(10-x)\,\text{cm}^3$ of $0.1\,\text{mol dm}^{-3}\ AgNO_3$ are equivalent to $\dfrac{35.5}{10} \times \dfrac{1}{1000} \times (10-x)\,\text{g}$ chloride. This is the amount in $2.0\,\text{cm}^3$ of urine. Convert your answer to g dm^{-3}.

Questions

a *Explain the chloride content of the two specimens of urine in terms of the regulation, by the kidney, of the salt content of the body.*

b *If you carried out a determination of chloride ions on blood collected at the same times as the urine specimens, what results would you expect?*

This investigation is based on:

ROBERTS, M. B. V. *Biology. A functional approach. Students' manual.* Nelson, 1974.